Security in ports

ILO and IMO code of practice

Security in ports

International Labour Office Geneva
International Maritime Organization London

ILO

Security in ports. ILO and IMO code of practice. Geneva, International Labour Office/London, International Maritime Organization, 2004

Safety, occupational safety, port, docker, code of practice. 15.06

ISBN 92-2-115286-3

Also published in French: *Sureté dans les ports. Recueil de directives pratiques du BIT et de l'OMI* (ISBN 92-2-215286-7, Geneva, 2004); and in Spanish: *Proteccion en los puertos. Repertorio de recomendaciones practicas de la OIT y de la OIM* (ISBN 92-2-315286-0, Geneva, 2004).

ILO Cataloguing in Publication Data

Cover page: Port of Las Palmas de Gran Canaria
Photocomposed by the ILO, Geneva, Switzerland
Printed in France

DTP
NOU

Preface

The Conference of Contracting Governments to the International Convention for the Safety of Life at Sea, 1974 (SOLAS Conference, London, 9-13 December 2002), adopted amendments to the International Convention for the Safety of Life at Sea (SOLAS), 1974. Amendments to SOLAS include a new Chapter XI-2 on special measures to enhance maritime security. This chapter is supplemented by the *International ship and port facility security code* (ISPS code), which contains, inter alia, requirements that relate to the security of the ship and to the immediate ship/port interface. The overall security of port areas was left to further joint work between the International Labour Organization (ILO) and the International Maritime Organization (IMO). Resolution No. 8, adopted by the 2002 SOLAS Conference, entitled "Enhancement of security in cooperation with the International Labour Organization (seafarers' identity documents and work on the wider issue of port security)", required the two additional elements shown in brackets to be addressed. This code of practice is the product of this cooperation.

The International Labour Conference, at its 91st Session in June 2003, adopted the Seafarers' Identity Documents Convention (Revised), 2003 (No. 185). The Convention provides for a uniform and global identity document that will permit the positive verifiable identification of the seafarer.

The Governing Body of the ILO, at its 286th Session in March 2003, and the Maritime Safety Committee of the IMO at its 77th Session in May-June 2003, established a working group of interested parties to draft a code of practice relating to security in ports. This draft was completed by the joint IMO/ILO working group in July 2003. The ILO Governing Body also agreed that the output of this working group should be formalized at a meeting of experts to be held in 2003, and adopted at that meeting. The working group consisted of representatives from employers, workers and governments, along with other organizations with a proper interest in the subject. A draft text was circulated to member States for comments in October 2003, before the meeting of experts took place, and those comments were collated and summarized for the experts in December 2003.

The code was finalized and adopted by the tripartite Meeting of Experts in Geneva, 8-17 December 2003. The meeting was attended by experts appointed following consultations with governments (Australia, Brazil, Canada, China, Egypt, Germany, Nigeria, Panama, Philippines, Spain, United Kingdom, United States), and by experts nominated by the Employers' group and the Workers' group of the ILO Governing Body. Expert observers from other governments, and observers from a number of intergovernmental and non-governmental organizations also attended.

The ILO Governing Body, at its 289th Session (March 2004), and the IMO Maritime Safety Committee (May 2004), approved the publication of this code.

This code of practice is not a legally binding instrument and is not intended to replace national laws and regulations. It is not intended to affect the fundamental principles and rights of workers provided by ILO instruments, or the facilitation of the access of workers' organizations to ports, terminals and vessels.

The practical recommendations contained in this code have been designed to provide guidance to all those responsible for addressing the issue of security in ports. The code will assist in identifying the roles and responsibilities of governments, employers and workers. It provides a proactive approach to security in ports and follows, where practicable, the practice and principles identified in SOLAS Chapter XI-2 and the ISPS code.

The ILO wishes to record its grateful thanks to Peter Green for his comprehensive work in developing the basic framework upon which this code was further developed and finalized.

List of participants

Experts nominated by Governments:

Mr. John Kilner, Assistant Secretary, Maritime Security, Department of Transport and Regional Services, Canberra (Australia).

Mr. John Platts, Special Advisor, Marine Security, Transport Canada, Security and Emergency and Preparedness Directorate, Ottawa (Canada).

Adviser:

Ms. Lynn Young, Director, Human Resources Development Canada (HRDC), Ottawa.

Mr. Daltro D'Arisbo, Labour Office Auditor – FISCAL, Porto Alegre (Brazil).

Adviser:

Captain Darlei Pinheiro, Mission Officer, Brazilian Permanent Representation to the IMO, London.

Mr. Ye Hongjun, Division Chief, Department of Worker Transport Administration, Ministry of Communication, Beijing (China).

Advisers:

Mr. Xu Yi, Director, Department of Human Resources, Ministry of Communication, Beijing.

Ms. Zhao Xiaoliang, Official, Department of International Cooperation, Ministry of Communication, Beijing.

Mr. Tarek Hassan Ibrahim Sharef Eldin, Director, Industrial Health and Safety Institute, Cairo (Egypt).

Advisers:

Mr. Hazem Abdel Hazem Halim, Head of Maritime Sector Administration, Alexandria.

Ms. Nadia El-Gazzar, Labour Counsellor, Permanent Mission of Egypt, Geneva.

Mr. Achim Sieker, Junior Unit Leader, Federal Ministry of Economics and Labour, Bonn (Germany).

Adviser:

Ms. Ute Bödecker, Executive for Legal and Security Matters, Ministry of the Interior, Hamburg.

Mr. Mobolaji Olurotimi Banjo, Acting Director, Inspectorate Department, Federal Ministry of Labour and Productivity, Abuja (Nigeria).

Advisers:

Mr. Wali Mansoor Kurawa, Director of Maritime Services, Federal Ministry of Transport, Abuja.

Mr. Sotonye Inyeinengi-Etomi, Special Assistant to the Honourable Minister of Transport, Federal Ministry of Transport, Abuja.

Dr. O.C. Nathaniel, Deputy Director, Joint Maritime Labour Industrial Council, Lagos.

Ms. Ifeoma Christina Nwankwo, Deputy Director, Federal Ministry of Labour and Productivity, Abuja.

Mr. John Idakwoji, Assistant Chief Administrative Officer, Abuja.

Mr. Audu Igho, Administrative Officer, Joint Maritime Labour Industrial Council, Lagos.

Mr. Henry A. Ajetunmobi, Assistant General Manager Security, Nigerian Ports Authority, Lagos.

Ms. Julissa Tejada de Humphrey, Director, Oficina Institu-cional Recursos Humanos, Ministerio de Trabajo y Desarrollo Laboral, Panama (Panama).

Mr. Gerardo S. Gatchalian, Supervising Labor and Employment Officer, Bureau of Working Conditions, Department of Labor and Employment, Manila (Philippines).

Advisers:

Mr. Benjamin B. Cecilio, Assistant General Manager for Operations, Philippine Ports Authority, Manila.

Ms. Yolanda C. Porschwitz, Labor Attaché, Permanent Mission of the Philippines, Geneva.

Mr. D. Juan Ramón Bres, Inspector, Inspección de Trabajo y Seguridad Social, Cadiz (Spain).

Adviser:

Mr. Pedro J. Roman Nuñez, Safety and Security Department Manager, Spanish Ports Administration, Madrid.

Mr. Graeme Henderson, Head of Marine and Civil Contingencies Section, Health and Safety Executive, London (United Kingdom).

Advisers:

Mr. David Carter, Deputy Head and Health and Safety Executive, Transport Safety Division, Marine, Aviation and Civil Contingencies Section, London.

Mr. Ashley Reeve, Head of Maritime Security Operations, Department of Transport, London.

Mr. Charles Thomas Pope, Area Director, US Department of Labor Occupational Safety and Health Administration (OSHA), Norfolk (United States).

Advisers:

Captain Jon S. Helmick, USMS, Director, Logistics and Intermodal Transportation Program, United States Merchant Marine Academy, Kings Point.

Captain David L. Scott, Chief, Office of Operating and Environmental Standards, Washington, DC.

Mr. John W. Chamberlin, First Secretary, Permanent Mission of the United States of America, Chambésy, Switzerland.

Experts nominated by the Employers:

Mr. Abdou Ba, Director of the Port Manpower Bureau, Syndicats des entreprises de manutention des ports du Sénégal, Mole (Sénégal).

Mr. Francis Bertrand, Director of Human Resources and Legal Affairs, International Organisation of Employers (IOE), Nantes (France).

Mr. Guido Marcelo Bocchio Carbajal, Legal Superintendent, Southern Peru Copper Corporation, Lima (Peru).

Mr. Joseph J. Cox, President, Chamber of Shipping of America, Washington, DC (United States).

Ms. Lynne Harrison, Human Resources Manager, Port of Napier Ltd., Napier (New Zealand).

Mr. Dierk Lindemann, Managing Director, German Shipowners' Association, Hamburg (Germany).

Mr. Claes Olmarker, Port Security Officer, Port of Göthenburg, Göthenburg (Sweden).

Mr. Usman Husein Punjwani, Seaboard Services Partner, Seaboard Services, Karachi (Pakistan).

Mr. Dahari Ujud, Senior Manager, Ancilliary Services, Port Klang (Malaysia).

Mr. Michael Joseph Van der Meer, General Manager, Port Authority Division, Namibian Ports Authority, Walvis Bay (Namibia).

Mr. Pieter M. Van der Sluis, Adviser Social Affairs, General Employers' Organization, Nieuwerherh (The Netherlands).

Adviser:

Mr. Fer M.J. Van de Laar, Chairman, International Association of Ports and Harbours, Environment and Marine Operations Committee, Amsterdam Port Authority, Amsterdam.

Mr. Alexander Zaitsev, President, Association of Ports and Shipowners of River Transport, Moscow (Russian Federation).

Experts nominated by the Workers:

Mr. Gary Brown, Port Security, AFL-CIO, Fife, Washington State (United States).

Mr. Marcel Carlstedt, Swedish Transport Workers' Union, Stockholm (Sweden).

Mr. P.M. Mohamed Haneef, Working President, Cochin Port Staff Association, Kochi (India).

Mr. Knud Hansen, Secretary, General Workers' Union, Copenhagen (Denmark).

Mr. Albert François Le Monnier, Third Vice-President, Safety Coordinator, International Longshore and Warehouse Union (ILWU), Vancouver (Canada).

Mr. Peter Lovkvist, Co-Sweden, Valbo (Sweden).

Mr. Kees Marges, Union Adviser, FNV Bondgeroter Netherlands, London (United Kingdom).

Ms. Veronica Mesatywa, National Sector Coordinator, Maritime Industry, South African Transport and Allied Workers' Union (SATAWU), Johannesburg (South Africa).

Mr. Leal Sundet, International Longshore and Warehouse Union (ILWU), California (United States).

Mr. James Trevor Tannock, Deputy National Secretary, Maritime Union of Australia, Sydney (Australia).

Mr. Kenji Yasuda, President, National Council of Dockworkers' Union of Japan (ZENKOKU-KOWAN), Tokyo (Japan).

Advisers:

Mr. Yuji Iijima, Chief of European Office, All Japan Seamen's Union, London.

Mr. Shimpei Suzuki, General Secretary, National Council of Dockworkers' Union of Japan, Tokyo.

In addition to the experts, Government observers from three member States, as well as ten representatives of intergovernmental and non-governmental organizations, attended.

Government observers:

Mr. Harri Halme, Senior Safety Officer, Ministry of Social Affairs and Health, Tampere (Finland).

Mr. Mario Alvino, Official, Ministry of Labour and Social Policy, Rome (Italy).

Ms. Asiye Türker, Senior Engineer, Head of Delegation, The Prime Ministry Undersecretariat for Maritime Affairs, Ankara (Turkey).

Representatives of intergovernmental organizations:

Mr. W. Elsner, Head of Port Unit, European Union (EU), Brussels.

Mr. Diego Teurelincx, Port Unit, EU, Brussels.

Mr. Christopher C. Trelawny, Senior Technical Officer, Navigational Safety and Maritime and Security Section, Maritime Safety Division, International Maritime Organization (IMO), London.

Mr. Viatcheslav Novikov, Economic Affairs Officer, United Nations Economic Commission for Europe (UNECE), Geneva.

Representatives of non-governmental international organizations:

Mr. Fer M.J. Van de Laar, Chairman, International Association of Ports and Harbours (IAPH), Environment and Marine Operations Committee, Amsterdam Port Authority, Amsterdam.

Mr. John Nicholls, Director, ICHA International Ltd., Romford, Essex.

Mr. Brian Parkinson, Adviser, International Shipping Federation (ISF), London.

Mr. Dan Cunniah, Director, Geneva Office, International Confederation of Free Trade Unions (ICFTU), Geneva.

Ms. Anna Biondi, Assistant Director, Geneva Office, ICFTU, Geneva.

Mr. Jean Dejardin, Adviser, International Organisation of Employers (IOE), Geneva.

ILO representatives:

Ms. Cleopatra Doumbia-Henry, Director, Sectoral Activities Department.

Mr. Dani Appave, Maritime Specialist, Maritime Activities, Sectoral Activities Department.

Mr. Marios Meletiou, Technical Specialist (Ports and Transport), Maritime Activities, Sectoral Activities Department.

Contents

Abbreviations

ILO	International Labour Organization
IMO	International Maritime Organization
ISPS	international ship and port facility security
PFSO	port facility security officer
PFSP	port facility security plan
PSA	port security assessment
PSAC	port security advisory committee
PSO	port security officer
PSP	port security plan
PT	potential target
SOLAS	International Convention for the Safety of Life at Sea, 1974, as amended
TRAM	threat and risk analysis matrix

1. Introduction

1.1. The objective of this code of practice on security in ports is to enable governments, employers, workers and other stakeholders to reduce the risk to ports from the threat posed by unlawful acts. The code provides a guidance framework to develop and implement a port security strategy appropriate to identified threats to security.

1.2. The code of practice on security in ports is part of an integrated approach to port-related security, safety and health issues, where security fits into existing health and safety guidance documents.

1.3. This code is intended to promote a common approach to port security amongst ILO and IMO member States.

1.4. This code is intended to be compatible with the provisions of SOLAS, the ISPS code and resolutions adopted by the 2002 SOLAS Conference. Where terms used in this code differ from those contained in the ISPS code, they are specified.

1.5. This code is not intended to replace the ISPS code. It extends the consideration of port security beyond the area of the port facility into the whole port.

1.6. The measures proposed within this code will apply to the entire port, including port facilities, as defined in the ISPS code; however, they should not replace the security measures in place within the port facility. The port security assessment (PSA) and the port security plan (PSP) should take into account the security measures in place within the port facilities, paying specific attention to the relationship between each port facility and the rest of the port.

1.7. This code provides a method of identifying potential weaknesses in a port's security and outlines security roles, tasks and measures to deter, detect and respond to unlawful acts against ports serving international traffic and maritime operations by:

1.7.1. Recommending that a security assessment is carried out by an appropriate authority in each port.

1.7.2. Recommending that a port security advisory committee (PSAC) be formed.

1.7.3. Recommending that a security plan be produced covering the issues identified in the assessment and identifying appropriate security measures to be implemented.

1.7.4. Applying security guidelines to all areas and functions of the port, and those working in, having business with and requiring access to the port or transiting through the port. This includes port workers and other port personnel, seafarers, passengers and passengers' baggage, cargo, material and stores, vehicles and equipment originating from within and outside the port area.

1.7.5. Promoting security awareness in the port and the training of personnel appropriate to their roles and responsibilities.

1.7.6. Maximizing the effectiveness of security measures through systematic drills, exercises, tests and audits of security procedures to identify and correct non-compliance, failures and weaknesses.

1.8. The port security guidelines in this code may also form a basis for security in domestic ports and maritime operations.

1.9. The code should be aligned with ILO and IMO member States' security and safety strategies. Nothing in this document is intended to prejudice the rights or obligations of States under international law, or to compromise the responsibility of national and local security organizations or other authorities and agencies to protect the safety and rights of people, property and operations within their area of jurisdiction.

1.10. This code is not intended to affect the fundamental principles and rights of workers provided by ILO instruments or the facilitation of the access of workers' organizations to ports, terminals and vessels.

1.11. This code does not affect obligations to comply with applicable national laws, regulations and rules.

2. Scope and definitions

2.1. *Scope*. This code of practice applies, as appropriate, to all persons, organizations or entities operating in, transiting through or having any other legitimate reason to be in the port.

2.2. *Definitions* used in this part of the code are, to the extent practicable, in keeping with those contained in the International Convention for the Safety of Life at Sea (SOLAS), 1974, as amended. For ease of reference, certain terms used in this code are defined in this section.

2.3. *Port.* For the purposes of this code, port means: "The geographic area defined by the member State or the designated authority, including port facilities as defined in the *International ship and port facility security (ISPS) code*, in which maritime and other activities occur."

2.4. *Designated authority.* The governmental organization(s) or the administration(s) identified within the member State responsible for the security of ports.

2.5. *Security.* A condition whereby the level of risk is deemed acceptable.

2.6. *Threat.* The likelihood that an unlawful act will be committed against a particular target, based on a perpetrator's intent and capability.

2.7. *Security incident.* Any act or circumstance affecting the security of a port.

2.8. *Security level.* The qualification of the degree of risk that a security incident will be attempted or will occur.

2.8.1. Security level 1 – The security level for which minimum appropriate protective security measures shall be maintained at all times.

2.8.2. Security level 2 – The security level for which appropriate additional protective security measures shall be maintained for a period of time as a result of heightened risk of a security incident.

2.8.3. Security level 3 – The security level for which further specific protective security measures shall be maintained for a limited period of time when a security incident is probable or imminent, although it may not be possible to identify the specific target.

2.9. *Port security officer* (*PSO*).[1] The person or persons tasked to manage and coordinate security in the port.

2.10. *Port security advisory committee* (*PSAC*). A committee established by the member State or the designated authority responsible, inter alia, to act as a security consultative body and to be involved in the continuous development and implementation of the port security plan.

2.11. *Port security assessment* (*PSA*). A comprehensive evaluation by the member State or the designated authority of threats, vulnerabilities, capabilities, preparedness and existing security measures related to a port, forming an essential and integral part of the process of developing a port security plan.

2.12. *Port security plan* (*PSP*). A written document that describes the measures that the member State or the designated authority and members of the port community should take to reduce vulnerabilities, deter threats and respond to security incidents. It should address issues impacting upon the security of the port and, where applicable, may

---

[1] The definition of port facility security officer (PFSO) is contained in Part A of the ISPS code, p. 8, item 2.1.8.

take into account issues relating to any port facility security plan or other security plan.

2.13. *Port facility*. A location as determined by the member State or by the designated authority where the ship/port interface, as described in the ISPS code, takes place.

2.14. *Infrastructure.* This is used in its broader meaning, which includes superstructures, services and other installations.

2.15. *Security-sensitive information.* Information, the disclosure of which would compromise the security of the port (including, but not limited to, information contained in any personnel-related file or privileged or confidential information that would compromise any person or organization).

3. Aim of security measures

3.1. The aim of port security measures is to maintain an acceptable level of risk at all security levels.

3.2. Security measures should be devised to reduce risks, and should, in the main, revolve around procedures to establish and control access to restricted areas and other vulnerable or sensitive key points, locations, functions or operations in the port.

3.3. Some examples of the aim of security measures that may be considered are to:

3.3.1. Prevent access to the port by persons without a legitimate reason to be there and prevent those persons with legitimate reasons to be in the port from gaining illegal access to ships or other restricted port areas for the purpose of committing unlawful acts.

3.3.2. Prevent introduction of unauthorized weapons, dangerous or hazardous substances and devices, into the port or vessels using the port.

3.3.3. Prevent personal injury or death, or damage to the port, port facility, ship or port infrastructure by explosive or other devices.

3.3.4. Prevent tampering with cargo, essential equipment, containers, utilities, protection systems, procedures and communications systems affecting the port.

3.3.5. Prevent smuggling of contraband, drugs, narcotics, other illegal substances and prohibited material.

3.3.6. Prevent other criminal activities, such as theft.

3.3.7. Protect against the unauthorized disclosure of classified material, commercially proprietary information or security-sensitive information.

4. Security policy

4.1. ILO and IMO member States should produce a "ports security policy statement" that provides the foundation to develop directives, rules and regulations as appropriate. Port security policies should take into account relevant international conventions, codes and other established national practices.

4.2. Member States should develop a security policy and ensure that a legal framework is in place to carry out the provisions of this code of practice. The security policy should address the following measures to be taken by the member States:

4.2.1. Promote regional and international cooperation.

4.2.2. Encourage maximum stakeholder participation in policy development.

4.2.3. Provide adequate resources to effectively implement and sustain security policy.

4.2.4. Recognize the importance of the human element: safety and security awareness, training and skill development.

4.2.5. Recognize the interdependence between security and public safety, economic development and protection of the environment.

4.3. The security policy should be periodically reviewed and updated to reflect changing circumstances.

5. Roles and tasks

5.1. *The ILO or IMO member State.* In addition to the development of a security policy, the member State should:

5.1.1. Identify the designated authority for each port required to have a PSP.

5.1.2. Ensure the establishment of a PSAC and the nomination of a PSO.

5.1.3. Nominate the persons responsible for port security operations in a specific port, as appropriate.

5.1.4. Ensure that a PSA is carried out.

5.1.5. Approve PSAs and any subsequent amendments thereto.

5.1.6. Ensure that PSPs are properly developed and implemented, and periodically reviewed and maintained.

5.1.7. Set and communicate the appropriate security level.

Member States may delegate any of the functions referred to in 5.1.2 through 5.1.6, above, to the designated authority.

5.2. *Port security officer* (PSO). Tasks of the PSO should include, inter alia, the following:

5.2.1. Conducting an initial comprehensive security survey of the port, taking into account the relevant PSA.

5.2.2. Ensuring the development and maintenance of the PSP.

5.2.3. Implementing the PSP.

5.2.4. Undertaking regular security inspections of the port, to ensure the continuation of appropriate measures.

5.2.5. Recommending and incorporating, as appropriate, modifications to the PSP in order to correct deficiencies and to update the plan to take into account relevant changes to the port.

5.2.6. Enhancing security awareness and vigilance of the port's personnel.

5.2.7. Ensuring that adequate training has been provided to personnel responsible for the security of the port.

5.2.8. Reporting to the relevant authorities and maintaining records of security incidents that affect the security of the port.

5.2.9. Coordinating implementation of the PSP with the appropriate persons or organizations.

5.2.10. Coordinating with security services, as appropriate.

5.2.11. Ensuring that standards for personnel responsible for security of the port are met.

5.2.12. Ensuring that security equipment is properly operated, tested, calibrated and maintained.

5.3. *Port security advisory committee* (PSAC). A PSAC should be established for every port, where applicable, with full terms of reference. It should act as a consultative and advisory body with a designated chairperson, and should cooperate with applicable safety and health committees, as appropriate. The PSAC's role should be to (as appropriate, but not limited to):

5.3.1. Advise on the implementation of the PSP and assist in conducting the PSA.

5.3.2. Coordinate, communicate and facilitate implementation of the applicable security measures required by the PSP.

5.3.3. Provide feedback on the implementation, drills and exercises, testing, security training and periodic updates of the PSP.

5.3.4. Ensure that its membership reflects the operational functions of the port and includes, as appropriate:

5.3.4.1. The port security officer (PSO) and the port facility security officer (PFSO)(s).

5.3.4.2. National and local government border control authorities and security agencies.

5.3.4.3. Police and emergency services.[1]

5.3.4.4. Worker's representatives.

5.3.4.5. Ship operator representatives.

5.3.4.6. Representatives of commercial concerns and tenants.

5.3.4.7. Trade associations.

5.3.4.8. Other relevant parties.

[1] Throughout this text, when the term "workers' representatives" is used, it refers to Article 3 of the Workers' Representatives Convention, 1971 (No. 135), which reads as follows:

For the purpose of this Convention the term "workers' representatives" means persons who are recognized as such under national law or practice, whether they are: (a) trade union representatives, namely, representatives designated or elected by trade unions or by the members of such unions; or (b) elected representatives, namely, representatives who are freely elected by the workers of the undertaking in accordance with provisions of national laws or regulations or of collective agreements and whose functions do not include activities which are recognized as the exclusive prerogative of trade unions in the country concerned.

6. Security level

6.1. The appropriate security level is determined by ILO and IMO member States. The security measures to be adopted appropriate to the security level should be outlined in the PSP.

6.2. Changes in the security level should be quickly communicated to those with a need to know in response to a perceived or actual change in threat information.

6.3. In the event of a change in security level, the PSO should act in accordance with the PSP, and verify that the requirements of the PSP and any additional or special security procedures appropriate to the particular threat are put into place. For example:

6.3.1. *Security level 1* measures may include random screening of personnel, baggage, material and stores, and vehicles, and the implementation of access and movement control.

6.3.2. *Security level 2* measures may include increased frequency of screening, more robust monitoring of the port, and more stringent access and movement control measures.

6.3.3. *Security level 3* measures may include 100 per cent screening, increased identification checks, temporary cessation of certain port activities and/or imposition of vessel traffic control measures, restricting access to certain areas, deployment of security personnel to key infrastructure, etc.

7. The port security assessment

7.1. The port security assessment (PSA) should be carried out by persons with the appropriate skills and should include the following:

7.1.1. Identification and evaluation of critical assets and infrastructure that it is important to protect.

7.1.2. Identification of threats to assets and infrastructure in order to establish and prioritize security measures.

7.1.3. Identification, selection and prioritization of measures and procedural changes and their level of acceptance in reducing vulnerability.

7.1.4. Identification of weaknesses, including human factors, in the infrastructure, policies and procedures.

7.1.5. Identification of perimeter protection, access control and personnel clearance requirements for access to restricted areas of the port.

7.1.6. Identification of the port perimeter and, where appropriate, identification of measures to control access to the port at various security levels.

7.1.7. Identification of the nature of the expected traffic into or out of the port (e.g. passengers, crew, ship/cargo type).

7.2. One example of a risk-based method and tool to assist in preparing a port security assessment is included in Appendix A. Other tools may be used.

8. The port security plan

8.1. The port security plan (PSP) should be based on the PSA and include:

8.1.1. Details of the security organization of the port.

8.1.2. Details of the port's links with other relevant authorities and the necessary communications systems to allow the effective continuous operation of the organization and its links with others.

8.1.3. Details of security level 1 measures, both operational and physical, that will be in place.

8.1.4. Details of the additional security measures that will allow the port to progress without delay to security level 2 and, when necessary, to security level 3.

8.1.5. Provision for the regular review, or audit, of the PSP and for its amendment in response to experience or changing circumstances.

8.1.6. Details of the reporting procedures to the appropriate member States' contact points.

8.1.7. Details of the necessary liaison and coordination between the PSO and any PFSOs.

8.1.8. Identification of restricted areas and measures to protect them at different security levels.

8.1.9. Procedures for the verification of identity documents.

8.1.10. Requirements for drills and exercises carried out at appropriate intervals to ensure the effective implementation of the PSP.

8.2. The PSP should refer to, and take into account, any other existing port emergency plan or other security plans.

8.3. The PSP should be protected from unauthorized access or disclosure.

8.4. An example layout and content of a port security plan is included in Appendix B.

9. Physical security of the port

9.1. At each security level, the PSP should identify the location of restricted areas, key points, vulnerable areas and critical functions in or associated with the port, and the physical protection and access control procedures and access documents required to reduce the level of risk.

9.2. Areas designated as "restricted areas" in the PSP should be delineated as such with appropriate warning signs, markings and, as appropriate to the security level in force, barriers and access control points.

9.3. Access control procedures should be established for restricted areas of the port for any person, vehicle, vessel, cargo, material, equipment and stores inbound or outbound whether from adjacent property or waterway, or from outside the port.

9.4. The PSP should define the procedures for:

9.4.1. The issuance, verification and return of access documents, at no cost to the workers.

9.4.2. The details of verification to be made regarding those persons required to be provided with, or issued, access documents.

9.4.3. The appropriate authorized access control requirements for each restricted area and level of access.

9.4.4. The reporting of lost, missing or stolen documents.

9.4.5. Dealing with the misuse of access documents.

These procedures should also cover temporary personnel, contractors and visitors at each security level.

9.5 The seafarers' identification document, issued in accordance with the Seafarers' Identity Documents Convention (Revised), 2003 (No. 185), would meet all requirements of this code of practice for the purposes of identification and access.

9.6. Where it is necessary to combine security aspects of the PSP and the PFSP, then these should be clearly identified in the PSP. These procedures should ensure that the security requirements are compliant with national and international customs and export regulations.

10. Security awareness and training

10.1. Security awareness is vital to the safety, security and health of port personnel and others having a place of work in the port, who should be made aware of their responsibilities to fellow workers, the port community and the environment. Appropriate training of personnel working in the port should maximize personal awareness of suspicious behaviour, incidents, events or objects when going about daily tasks, and the invaluable contribution to be made to the security of the port and its personnel by each individual. Clear lines for reporting such matters to supervisors, managers or appropriate authorities should be included. Additional or special training may be required for people in particular roles.

10.2. Training may be focused on particular roles and tasks in the port or at external facilities serving the port such as:

10.2.1. Security and law enforcement personnel.

10.2.2. Stevedores and all those handling, storing and transporting, or coming into contact with, passengers, freight, cargo, material and stores or ships.

10.2.3. Other associated roles and tasks where personnel do not come into direct contact with passengers, freight, cargo, material and stores or ships as a matter of course, but who are in administrative and support roles in the port or at associated facilities.

10.3. Consideration should also be given to circumstances where it would be ineffective or contrary to good security practice to train or give additional information to those without a direct need to know.

11. Confidentiality and non-disclosure of information

Contracts of employment or organizational rules should contain provisions requiring personnel not to divulge security-related information on the port, security training, access control systems, locations of security or communications equipment and routines, or business of the port to persons who do not have a direct need to know.

Appendix A

The port security assessment

Introduction

1. The threat and risk analysis matrix (TRAM) is a simplified risk-based method and tool to assist in carrying out a port security assessment (PSA). It is only one of a number of tools and is given here by way of example.

2. Its purpose is to identify threats with a view to initiating and recommending countermeasures to deter, detect and reduce the consequences of any potential incident, should it occur. Such an analysis may be a valuable aid to allocation of resources, forward planning, contingency planning and budgeting.

3. The TRAM should be updated as often as changing circumstances may dictate to maintain its effectiveness. This task would normally fall under the remit of the designated authority, which should establish and maintain close links with security committees, and key commercial and industrial service partners and customers.

4. In addition to the more obvious threats, the list of potential targets should be as comprehensive as possible with due regard to the function(s) of the port, legal, political, social, geographic and economic environment of the country, and the security environment specific to the port.

Assessment process

5. Table 1 is a blank version of the TRAM. The object is to compare/evaluate security measures that will reduce, independently, the vulnerability or impact and collectively reduce the overall risk score. It should be borne in mind that introducing a security measure for one threat may increase the risk of another.

6. *Potential target (PT).* (There should be a separate table for each potential target.) Identify PTs through assessment

Table 1. Blank threat and risk analysis matrix (TRAM)

Potential target: Person/place/location (identify each PT in the port area not covered by the PFSP or other official subordinate plan)

Scenario No.	Threat scenario	Threat	Vulnerability	Impact	Risk score	Action priority
A	**B**	**C**	**D**	**E**	**F**	**G**
1						
2						
3						
4						
5						
6						
7						
8						
9						

of functions and operations, vulnerable areas, key points or persons in the port and in the immediate environs that may, if subject to an unlawful act, detrimentally impact on the security, safety of personnel or function of the port.

6.1. Establish "ownership" of the identified PT. For example:

6.1.1. Directly owned and controlled by the port operator or member State.

6.1.2. Directly owned by the port operator or member State but rented, leased, occupied and controlled by other parties.

6.1.3. Owned, controlled and operated by other parties.

6.1.3.1. Represented on the PSAC.

6.1.3.2. Not represented on the PSAC (consider whether membership would be appropriate and/or beneficial to the port community).

7. Establish if there are any existing security measures, such as a perimeter fence, access control and/or security patrol or monitoring of the PT. If so, are they effective, or can improvements be made?

8. *Threat scenario* (columns A and B of table 1). Consider threat scenarios from both internal and external sources to which the identified PT may be vulnerable (input from police, security and intelligence services is essential).

8.1. Threat scenarios (among many) that it may be appropriate to consider:

8.1.1. Direct attack to cause injury and loss of life, or destroy functions and infrastructure of the port. To take over vehicles/vessels as means to inflict damage by ramming. Release of noxious or hazardous material either from vehicles/vessels or storage areas, and so on.

8.1.2. Sabotage.

8.1.3. Kidnap and ransom (for reward, extortion or coercion).

9. *Threat* (column C of table 1). The probability of an incident occurring should be assessed on the following scale:

3 = High
2 = Medium
1 = Low

The allocation of a particular threat score may be based on specific information received or the known characteristics of the potential target.

10. *Vulnerability* (column D of table 1). The vulnerability of the PT to each threat may be assessed as follows:

4 = No existing security measures/existing security measures are not effective (e.g. unrestricted access to target; target not monitored; personnel untrained; target easily damaged).

3 = Minimal security measures (e.g. restricted areas not clearly identified; inadequate access control procedures; sporadic monitoring; no formal security training programme; target susceptible to certain types of damage).

2 = Satisfactory security measures (e.g. restricted areas clearly identified and access controlled; formal security training programme; adequate monitoring and threat awareness; target not easily damaged).

1 = Fully effective security measures (e.g. all of 2 above, plus: capable of promptly scaling to higher security level as needed; target difficult to damage or has sufficient redundancy to prevent disruption if certain functions are damaged).

11. *Impact* (column E of table 1). Assess the impact (consequence) of each potential incident on the PT and port should it occur. Specific "impacts" and priorities for a particular port may be substituted by the designated authority to meet the national security profile and requirements:

5 = Detrimental to security and safety (likely to cause loss of life, serious injuries and/or create widespread danger to public health and safety).

4 = Detrimental to public safety and/or national prestige (likely to cause significant environmental damage and/or localized public health and safety).

3 = Detrimental to the environment and/or economic function of the port (likely to cause sustained port-wide disruption and/or significant economic loss and/or damage to national prestige).

2 = Detrimental to assets, infrastructure, utility and cargo security (likely to cause limited disruption to an individual asset, infrastructure or organization).

1 = Detrimental to customer/port community confidence.

12. *Risk score* (column F of table 1). The score is the product of threat x vulnerability x impact.

12.1. The highest score scenario will be:

Threat – High	3
Vulnerability – No existing countermeasure	4
Impact – Potential loss of life/injury	5
Risk score	60

12.2. The lowest score scenario will be:

Threat – Low	1
Vulnerability – Fully compliant	1
Impact – Little	1
Risk score	1

13. *Action priority* (column G of table 1). Tabulating and listing the scores for each threat against each PT will assist in assessing the priority in which to deal with each potential incident. The process should lead to indications of actions required to deter, detect and mitigate the consequences of potential incidents, resources available or required, and appropriate security measures.

14. In assessing likely scenarios, the history and modus operandi of illegal groups most likely to operate in the area should be considered when identifying the PT, and determining and assessing the most appropriate security measures.

15. This assessed reduction of the score for each scenario is based on the perceived effectiveness of the security measures when they have been put into effect. The result should give some

guidance as to which actions and resources will have the greatest benefit in deterring attack of the PT. It may also indicate that some targets or threats do not need to be considered or that the security measure is not achievable because of resource or other constraints.

16. The TRAM for every potential target should be collated into one master matrix of similar threat scenarios and common security measures identified to give the maximum benefit. It may also be that some PT may be grouped together under one security measure. For example, one or more PTs close together may be contained within one perimeter fence with one gate controller. It may be that a vulnerable operation in a remote part of the port can be moved into a more secure area. Every possible realistic action should be considered.

17. The completed TRAM, together with a consolidated summary of all security measures that have been devised and are able to be implemented, should form the basis from which the port security plan can be developed.

Assessment example

The ten-step example on pages 27 to 36 is used to illustrate the possible working of a security assessment using the TRAM for a specific threat scenario – destroy port authority's communication tower by explosives.

Step 1 – List feasible scenario in column B

Scenario No.	Threat scenario	Threat	Vulnerability	Impact	Risk score	Action priority
A	**B**	**C**	**D**	**E**	**F**	**G**
1	Destroy port authority's communication tower by explosives					

Feasibility scenario as determined by current port security assessment	The tower is a critical component of port operational and commercial communications. It supports booster stations for local police and emergency service communications and, in addition, supports mobile telephone repeater services for the area. Currently the tower is protected from casual access or interference by a 2-metre high razor wire fence of 15 metre diameter, and is located in a non-restricted area approximately 200 metres from the Harbour Master's Office. The facility is positioned on flat ground approachable from all sides, and a service road, which is accessible from the public area roads, passes within 20 metres of the perimeter fence. Access to the compound is limited to maintenance and servicing of the tower components, as required, and seasonal ground maintenance, including grass cutting by regular port-approved contractors. There is a mobile security patrol that visits and checks for signs of damage or intrusion once by day and once by night. The tower could be easily damaged by an explosive device thrown over the fence or placed against the fence, or a car bomb driven up to the compound or placed on the service approach road.

Step 2 – Assign a threat score to this scenario in column C

Scenario No.	Threat scenario	Threat	Vulnerability	Impact	Risk score	Action priority
A	**B**	**C**	**D**	**E**	**F**	**G**
1	Destroy port authority's communication tower by explosives	1				

Threat score based on intelligence, security level, current deterrent measures and other relevant factors	This scenario has been given a threat score of 1 (low) because no specific intelligence has been obtained that suggests communications facilities are being targeted at the present time. A score of 2 (medium) or 3 (high) may be given based upon intelligence.

Step 3 – Assign a vulnerability score to this scenario in column D

Scenario No.	Threat scenario	Threat	Vulnerability	Impact	Risk score	Action priority
A	**B**	**C**	**D**	**E**	**F**	**G**
1	Destroy port authority's communication tower by explosives	1	2			

Vulnerability is the susceptibility of a potential target to a particular threat

In this example, the threat is damage to the communication tower by explosives. Vulnerability is listed as 2 (satisfactory security measures) because the facility's existing perimeter fence and security patrol are considered a sufficient deterrent.

Step 4 – Assign impact score to this scenario in column E

Scenario No.	Threat scenario	Threat	Vulnerability	Impact	Risk score	Action priority
A	**B**	**C**	**D**	**E**	**F**	**G**
1	Destroy port authority's communication tower by explosives	1	2	3		

Impact is the consequence of an incident – the effect on public health, safety or security, etc.

In this example, impact is listed as 3 (detrimental to the economic function of the port) because there is no back-up communication tower, so its loss would shut down the port for some time until repairs could be made, thus causing substantial economic loss. Impact may be further reduced if there is redundancy to the potential target (e.g. a back-up communication tower) or if a target may be easily repaired. Conversely, impact may increase if there is no redundancy, or if a target would be difficult to replace.

Step 5 – Calculate the initial risk score in column F

Scenario No.	Threat scenario	Threat	Vulnerability	Impact	Risk score	Action priority
A	**B**	**C**	**D**	**E**	**F**	**G**
1	Destroy port authority's communication tower by explosives	1	2	3	6	

The initial score is calculated by multiplying columns C, D and E

In this example, the initial score would be 6 (1 x 2 x 3 = 6).

Step 6 – Determine the action priority in column G (typically performed following several scenario calculations)

Scenario No.	Threat scenario	Threat	Vulnerability	Impact	Risk score	Action priority
A	**B**	**C**	**D**	**E**	**F**	**G**
1	Destroy port authority's communication tower by explosives	1	2	3	6	

The action priority is based on each scenario's initial score

Establishing action priorities based on initial risk scores is a quick way to distinguish between the various scenarios, and can help focus and allocate scarce resources, particularly when a large number of scenarios are assessed.

Step 7 – Determine new scores and action priorities based on changes to threat, vulnerability or impact

Scenario No.	Threat scenario	Threat	Vulnerability	Impact	Risk score	Action priority
A	**B**	**C**	**D**	**E**	**F**	**G**
1	Destroy port authority's communication tower by explosives	2	2	3	12	

A variety of factors may change an initial risk score

For example, an increase in threat from 1 (low) to 2 (medium) would raise the risk score from 6 to 12 (column C increases from 1 to 2, thus 2 x 2 x 3 = 12; see above). When the threat score increases, persons involved in developing security measures can use this table to recalculate how vulnerability reduction measures or impact reduction measures may reduce the risk score. If 6 is deemed to be an acceptable level, then vulnerability reduction measures or impact reduction measures should be considered that will reduce the figures in columns D and E so as to give a risk score in column F of no higher than 6.

Step 8 – Implementing measures to reduce vulnerability

As described in Step 1, the tower is protected from casual access or interference by a 2-metre high razor wire fence of 15 metre diameter and is located in a non-restricted area approximately 200 metres from the Harbour Master's Office. The facility is positioned on flat ground approachable from all sides, and a service road, which is accessible from the public area roads, passes within 20 metres of the perimeter fence. Access to the compound is limited to maintenance and servicing of the tower components, as required, and seasonal ground maintenance, including grass cutting by regular port-approved contractors. There is a mobile security patrol that visits and checks for signs of damage or intrusion once by day and once by night. With these measures, vulnerability was scored as 2. However, if additional vulnerability reduction measures, such as a full-time on-site security force, or changing the non-restricted area to a restricted area, the vulnerability score may be reduced from 2 to 1, fully effective security measures. Thus, with vulnerability in column D reduced from 2 to 1, as shown below, a new risk score of 6 is produced.

Scenario No.	Threat scenario	Threat	Vulnerability	Impact	Risk score	Action priority
A	**B**	**C**	**D**	**E**	**F**	**G**
1	Destroy port authority's communication tower by explosives	2	1	3	6	

Step 9 – Implementing measures to reduce impact

Reducing the impact will alter the figure in column E and reduce the overall risk score. Recall that the tower is a critical component of port operational and commercial communications. It also supports booster stations for local police and emergency service communications. In addition, the mast supports mobile telephone repeater services for the area. Assuming that there is no back-up communication tower, the impact of losing this tower was initially calculated as 3 in column E. However, if a back-up facility were available, it would create some redundancy, thereby reducing the impact of a loss. Thus, with impact reduced from 3 to 2, limited disruption to port organization due to communications redundancy, as shown below, produces a new risk score of 8. While this is an improvement from 12, the persons responsible for port security could then decide whether additional measures were needed.

Scenario No.	Threat scenario	Threat	Vulnerability	Impact	Risk score	Action priority
A	**B**	**C**	**D**	**E**	**F**	**G**
1	Destroy port authority's communication tower by explosives	2	2	2	8	

Step 10 – Implementing measures to reduce vulnerability and impact

If both the vulnerability reduction measures and impact reduction measures discussed in this example were taken together, the total risk score would be reduced to 4, well below the initial score of 6.

Scenario No.	Threat scenario	Threat	Vulnerability	Impact	Risk score	Action priority
A	**B**	**C**	**D**	**E**	**F**	**G**
1	Destroy port authority's communication tower by explosives	2	1	2	4	

The persons doing the security assessment and persons charged with implementing security measures must determine the effectiveness of various vulnerability or impact reduction measures for their ports.

Appendix B

The port security plan

Introduction

1. The port security plan (PSP) should be compatible with the ISPS code, ship security plan and port facility security plans.

2. It is not intended that the PSP should duplicate or replace the PFSP. It may however identify the relationship with the port facility and provide the transition of maritime security from the ship through the port facility into and from the port.

3. The PSP should address at least the following for each security level:

3.1. Identification of the person in the port designated to receive security-sensitive information affecting the port.

3.2. Measures designed to prevent unauthorized weapons or any dangerous substances and devices intended for use against persons, port assets or infrastructure and facilities from being introduced into the port.

3.3. Identification of restricted areas of the port and measures designed to prevent unauthorized access.

3.4. Procedures for responding to security threats to the port or breaches of security, including provisions to maintain critical operations of the port.

3.5. Approval of the security measures required at each security level and, in particular, procedures for responding to any security instructions the member State may give at security level 3.

3.6. Procedures for evacuation in the event of a security incident.

3.7. Duties of port personnel assigned security responsibilities.

3.8. Procedures for interfacing with port facility security activities.

3.9. Procedures for periodic review and updating of the PSP.

3.10. Procedures for reporting security incidents.

3.11. Identification of the PSO and 24-hour contact details.

3.12. Measures to ensure the security of information contained in the plan including, where appropriate, proprietary information of members of the port community.

3.13. Measures to prevent interference or theft of port property and equipment, and inbound and outbound material, stores and cargo.

3.14. Procedures for auditing the PSP.

3.15. Procedures for responding to security alarm activation at the port facility or other restricted area in the port.

3.16. Procedures to facilitate the movement and access of seafarers, including representatives of seafarer welfare organizations and workers' organizations, to the port, port facility and ship as appropriate.

4. The PSP may be used in addition to identify and communicate:

4.1. Permitted inbound and outbound passenger routing.

4.2. Inbound and outbound seafarer routing (from/to port facility/between port facilities).

4.3. Holding areas and routing for inbound and outbound (inter-ship/port facility) and transiting cargo, materials, stores and traffic.

4.4. Approved holding areas for dangerous goods and hazardous material.

4.5. The form of the physical interface with the port facility (facilities) and movement of persons, material, stores and cargo between port facilities.

4.6. Safe and secure routes to, and area for holding, suspect explosive devices and other suspicious objects.

Roles and tasks

5. The designated authority should require that all ports devise a PSP and nominate a PSO who along with the PSAC should implement the plan.

Format and content of the PSP

6. By way of example, the following is given to assist the production of the PSP that may be made up of or contain the following information.

7. *Front/header page*
   - Name of port area
   - List of associated plans
   - List of members of PSAC
   - Name, appointment and signature of person approving the plan
   - Date of approval
   - Authority for issue
   - Date of issue

8. *Distribution list* – for unclassified and classified parts of the plan.

9. *Record of changes* – explanation of change procedures and tasks of plan holders to amend the plan and implement changes.

10. *Table of contents* – appendices may be used to segregate classified or commercially sensitive information and only distributed to those members of the port community approved to receive the information.

11. *Introduction.* An explanation of the background, circumstances and objective of the PSP. Include major objectives and security policies, e.g. to deter, detect and respond through promotion of a high level of security awareness and training.

12. *Security policy statement.* Include a statement of the port security policy.

13. *Assumptions,* e.g.:

13.1. Unlawful acts may occur at any time with little or no warning.

13.2. Protection of human life, health and security is the most important consideration in development of the plan.

13.3. Maintaining the free flow of commerce and the functioning of the port is a critical consideration.

13.4. No single entity can provide all the resources required to provide adequate security measures and response to the consequences of an unlawful act.

13.5. Other disaster and contingency plans (e.g. dangerous goods, hazardous material or natural disaster response) will be activated as appropriate in response to any security incident.

13.6. All members of the port community will voluntarily support and participate in measures to secure the port and its functions.

14. *Port security advisory committee* charter, if applicable, or authority for the committee's formation and:

14.1. Brief of role and task of the PSAC, e.g.:

14.1.1. To consult and advise on the implementation of the PSP and other security matters, as appropriate.

14.1.2. To develop procedures for sharing and communication of security-related information.

14.1.3. To promote security awareness as a deterrent to unlawful acts.

15. *Organization and membership of the PSAC.* Composition of members of the PSAC, and the relationship with other port and national or local planning committees.

16. *The port.* Define the geographical and functional perimeter [boundaries] and make-up of the port, including all waterways and modes of transport, infrastructure and port and commercial functions.

17. Include associated infrastructure, facilities, functions and secondary ports to which a security threat may relate and that may be included in the main plan or other security plans.

18. List local law enforcement agencies and municipal emergency and support services (include local hospital/medical facilities) that may contribute to response and consequence management.

19. *Maps and charts.* Provide maps and charts showing all salient features and location of operations, functions, and routes, and access points including appropriate navigation channels. This may be attached as an annex to the plan.

20. *Operations and functions.* Detail maritime and non-maritime operations and functions.

21. *Critical operations and activities.* Identify and describe all critical operations and other significant activities carried out in the port area.

22. *Security levels*

22.1. *Security level 1.* The level at which minimum appropriate protective security measures shall be maintained at all times.

22.2. *Security level 2.* The level at which appropriate additional protective security measures shall be maintained for a period of time as a result of heightened risk of a security incident.

22.3. *Security level 3.* The level at which further specific protective security measures shall be maintained for a limited period of time when a security incident is probable or imminent, although it may not be possible to identify the specific target.

23. *Communications.* Describe and detail the means of communicating security level(s), changes to the security level and methods of raising alarm in the event of an incident.

24. *Security measures, procedures and operations.* Tabulate and list in detail all security measures and operations that are to be implemented in the port at each security level in response to issues identified in the security assessment.

25. This should cover personnel security, perimeter and physical barriers, access control and all approved security measures. It should detail the roles and tasks of all members of the port

community to establish/monitor/control, as appropriate, restricted areas and navigation zones.

26. It may be appropriate to use existing procedures to aid communication, implementation and testing. Where appropriate, functional operating procedures and working instructions are in existence, it may be feasible to add security elements to such procedures and working instructions. For example, if there is an existing written operational procedure for checking contents of inbound vehicles against other documentation or information, it may be possible to include security inspection of the contents in the existing procedure.

Roles, resources, authorities and tasks

27. Detail how and by whom security procedures will be implemented.

Relationship to other plans and organizations

28. List all other plans and organizations that may contribute to, relate to or impact on the PSP.

Response and crisis management

29. Identify and list agencies and contacts responsible for responding to mitigate the cause or consequence of an incident. Devise, tabulate and communicate a response plan for every perceived incident.

PSP review and maintenance policy

30. Define the policy and procedures to review and maintain the PSP.

PSP security and control

31. Define the distribution, dissemination and security of the plan, or parts of the plan, to achieve widest communication of its requirements without compromising security or proprietary information.

Training

32. Detail training requirements for port personnel to fulfil their role and that of their organization in carrying out tasks under the PSP.

Drills, exercises and testing

33. Methods should be detailed to carry out drills and exercises and to test the plan periodically, to check that it remains current and achievable by identifying changes that may impact on any critical response, resource or consequence factor.

Appendix C

Selected references

The information given in this appendix is intended to provide background and references on the code and other sources of information that may be of interest.

1. Details of the following may be found on the IMO web site – www.imo.org :

1.1. International Convention for the Safety of Life at Sea, 1974 (SOLAS) (as amended).

1.2. ISPS code, 2003 edition: *International ship and port facility security code and SOLAS amendments 2002* (London, IMO, 2003).

2. Seafarers' Identity Documents Convention (Revised), 2003 (No. 185) (available on the ILO web site – www.ilo.org).

3. Details of the following may be found on the United Nations web site: www.un.org/docs :

3.1. United Nations General Assembly resolution 57/219 (2002): Protecting human rights and fundamental freedoms while countering terrorism.

3.2. United Nations Security Council resolution 1373 (2001): Threats to international peace and security caused by terrorist acts.